EARTHQUAKE
AND
CASUALTIES

AUTHOR
BAHRAM   HOJABRI

IN THIS BOOK
A NEW THEORY FOR EARTHQUAKE
AND
HOW AND WHY IT MAY DEMOLISH
AND
NO REASON FOR PANIC
ARE
DESCRIBED

A
NEW
CONCEPT
ON
EARTHQUAKES
AND
RELATED SUBJECTS

Second Edition

PREFACE

Dear Readers,

This is a new theory about Earthquakes, which is very logical and simple to follow. It takes a different view from the norm and introduces new findings about earthquakes.

In this book, an earthquake is explained as to what it is and how it happens, so that people will be aware of the facts and this will help eliminate fear and panic, the same way that it was eliminated when the facts were found about thunderstorms.

Persons who are listed as missing may be injured but are alive and trapped under the rubble or may be dead so I believe, all missing persons should be found dead or alive.

Regardless of the causes of earthquakes, if the facts written in this book can help people to overcome their fears and panic due to earthquakes, more people can be saved.

I have done the share for my given life.
As English is not my prime language so I apologize for composition mistakes.

CONTENTS

CHAPTER

1

AN EARTHQUAKE

Part 1 An Earthquake in the Middle East

It was the summer of 1961, late at night when for the first time in my life I felt an earthquake. Since we were hundreds of kilometers from the epicenter, all we felt was the gentle movement of our building. At that point, all we knew was that it must have been a strong yet far away earthquake.

Later on, I was sent on a mission to that area, when I saw the destruction and the casualties, which made me realize how powerful the earthquake was.

Thousands of square kilometers of rugged land were severely shaken, and about thirty villages were destroyed.

Among those villages, there was one in which hundreds of people used to live, with many public buildings. What was left was some piles of dirt that used to be buildings and about 20 injured persons.

One of the reasons for such mass destruction, besides the high magnitude of the quake, is the material of which the buildings were made, in this case "mud", mixed with straw. Since dried mud has no elasticity, it does not hold well.

The quake happened about 200 kilometers west of Tehran, the capital city of Iran. It was a terrible disaster, over 4,000 people died, and the survivors were left injured and homeless.

Many countries contributed to the rebuilding of these villages, mainly the U.S.

As there were no roads, at that area battalion size helicopters were brought in to help transport the injured and deliver food to needy people in mountainous areas.

An army field hospital was set up which worked constantly to attend to the injured.

Being a witness to the mass destruction and the two inch gap on the ground which was extended for kilometers made me think about the source of energy, which must come from deep within the ground, that when is released, can shake such a vast rugged land. Since that time I have been studying many books in different languages, all saying the same thing but in different phrases.

Part 2 EARTHQUAKE IN LOS ANGELES

In recent years, many earthquakes have shook Los Angeles county, of which 2 were severe and remarkable.

The 1989 earthquake in L.A. left some casualties, and destruction in a limited area. But what was quite noticeable was the extremely loud sound similar to an explosion. This is the case which happens in many earthquakes. Another example is the earthquake that happened in Kobe, Japan in January of 1995, where a loud sound was heard when the quake hit.

The other earthquake happened in Northridge, January of 1994, a city in LA. county. There were about 50 casualties, and much destruction. Within a 30 mile radius, in addition to buildings, many multi level parking lots of shopping center were demolished.

The remarkable case is the nine thousand low magnitude earthquakes which happened over many months following the quake, called "After Shocks".

The focus of this earthquake was at a depth of ten miles underneath the ground.

Since an earthquake has Omni direction, we can imagine that about thirty thousand Cubic Miles of rugged ground was severely shaken.

A huge amount of energy is required to do this job.

So far, we have explained what happens during an earthquake, for those readers who have never experienced one. This will give them an idea, and help them to follow the upcoming discussions about earth-quake.

The known facts about earthquakes, up to this point, besides demolishing, are:

- The effects of an earthquake indicate that it is caused by a sudden release of energy.
- The released energy as registered is a point called Focus.
- Small magnitude earthquakes are often called After Shocks.
- Earthquakes can not be anticipated.

CHAPTER

2

EARTHQUAKE
PRESENT THEORY

Part 1 Earthquake Concept

To discuss the present theory of an earthquake, we should have a look at some geological concepts. And as we are not going to study geology, what is needed to know is as follows:

- Crust and plate.

  Continents and the oceans are set on a layer, called earth crust. This layer is set on a part called lithosphere, which has a depth of 50 to 100 kilometers. At some sections, lithosphere are broken, which is called plate. These broken parts or plates have a movement of about 1 inch every year, which is the continental drift concept.

- Fault or gap in the ground is called a "FAULT". It can have a length of less than an inch, or many miles long, which at some parts may be disconnected.

- Present concept of earthquake.

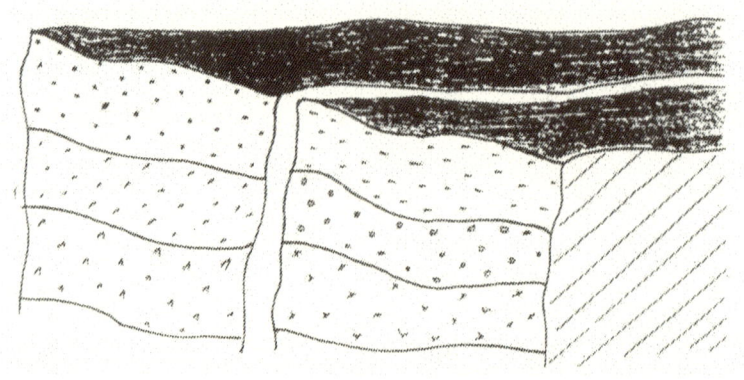

FIG. 1 SUPPOSED A PLATE EDGE APPROACHING OTHER EDGE.

It is believed, that if the movement of plate edge sliding on each other are stopped by friction between rocks, a stress or pressure will develop and build up to a point, beyond which, rocks will crash, and a sudden movement will happen, consequently, the pressure will be released and rocks will take a new position, which are not under stress.

This sudden movement between rocks to release pressure is what will cause an earthquake.

In some cases, it is also stated that the release of pressure between faults can be the cause of an earthquake.

Part 2 Uncleared Cases

A theory, describing a subject. If there are some uncleared cases, then the theory is not valid and is questionable.

Refer to Fig. 1 on page 8 which is more realistic than other figures for this subject.

Based on what is known about earthquakes, then the following are cases which aren't explained by the present theory of earthquakes.

- The plate edge, as described by geology, should be a huge wall, about 20 kilometers high, and thousands kilometers long. When two such surfaces slide on each other, then the friction will be between rocks the size of a mountain, so the released pressure will be a surface. Whereas an earthquake focus is registered as a spot.

- Fault is a superficial crack on the ground, it appears due to an earthquake, and never has a noticeable movement and no source of energy exists between the two surfaces. So how can it cause an earthquake.

- How could the released pressure between rocks, which happens now, make more than thousands of after shocks, over a many months following the earthquake.

- It was announced that an earthquake happened at a depth of 350 miles underneath the ground. Since these are not fault or plate at that depth, then why is it called an earthquake, not an explosion?

- The waves which are called Body Wave (P) and Sheer Wave (S) are not known to science. A wave has specific characteristics, as wavelength and a source to generate. So to which of these categories do these waves belong. In some books, (S) waves are said to be surface waves, which is not correct.

- The release of pressure between rocks is in one direction, and in the first few seconds of release, pressure should drop to zero. But earthquakes shakes moves back and forth, and in some cases last for some minutes. So what is the cause of vibration and duration of earthquakes.

- Earthquake location and plate edge. Fig. 2 on page 14 shows the approximate locations of the earthquakes that happened during a few months around the cities of L.A. county.

What could be the cause of the three earthquakes which happened within 4 months in San Fernando city and 2 earthquakes that happened in a day at the city of Dana Point, in respect to the 1 inch movement per year.

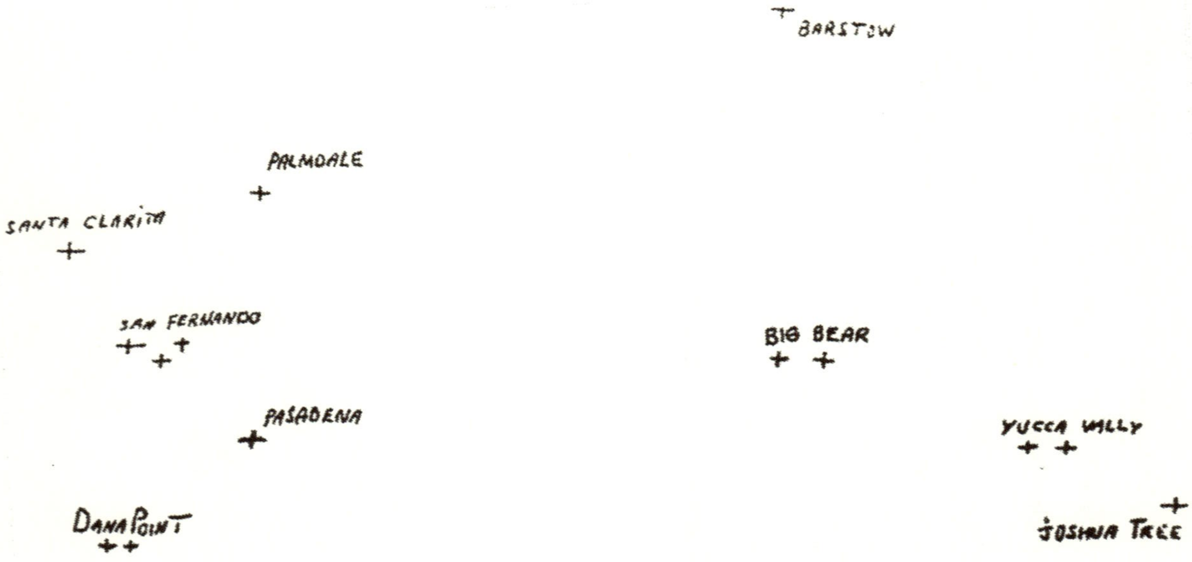

FIG. 2 APROXMATE LOCATIONS OF SOME EARTHQUAKE AT AN AREA OF 20 BY 40 MILES.

3

NEW THEORY
ON EARTHQUAKES

Part 1 Earthquake's Basics

When reviewing the present concept of earthquakes, and the unclear cases, it seems to not have much chance for new findings more than we already know. But by putting together different subjects of science, and comparing them, to what is known about earthquake then we will be able to come up with a new concept about earthquakes, which is what I have attempted to do.

Two cases are obvious about earthquakes, one that it occurs deep underneath the earth. Second that since an earthquake is the same as doing "work" it needs some sort of an energy.

Therefore there must be a source of energy that when released, can cause an earthquake.

An earthquake starts at a point, called "Focus", and based on and indications shows that it is , caused by an explosion. So let us find out if an explosion can happen underneath the ground.

Part 2 How Explosions Happen

Before trying to find sources of energy deep underneath the ground, in between the different layers, let us see how an explosion occurs, and what it needs. For those of our dear readers who are familiar with demolition, it is evident that the following three factors are needed for an explosion to happen.

- Explosive material

  There are a number of different explosive materials, such as TNT, gun powder or fuel gas such as Methane, etc.

  Among the above mentioned, Methane gas can be found in nature, but the rest are man made. For Methane, or natural gas to explode, it needs the right amount of air, otherwise a pure gas with no oxygen will never burn or explode.

  In the car industry, by adjusting the carburetor, the right mixture of gas and oxygen (air) is made, (about 16%), which is called engine tune up.

- Demolition Pit

This is the location where explosive material is placed, to explode. It could be anywhere form a hole dug up in a coal mine wall, where dynamite is placed, or the chamber of a gun used to hold gun powder. The tightness of this pit or chamber, has a coefficient of explosion up to some thousands for explosion. For example, if some gun powder is ignited in the open air, it will burn like a piece of paper, whereas in a gun chamber it will explode.

The same is true for Methane gas. In a stove, gas burns, but if a closed room is filled with the same gas, then one spark can cause a massive explosion, capable of destroying parts of a building.

Accidents like these happen all the time.

- Ignition

In order to explode an explosive material, some means of ignition is necessary.

In case of Methane gas, in addition to need a certain amount of air, it can be ignited by two methods. The same methods that are used in the internal combustion engine.

In a diesel engine, by applying a high compression on the fuel gas in the cylinder, the temperature will rise to a degree where it will ignite the fuel gas and consequently an explosion will occur.

The second method to ignite Methane gas is by making some sparks, same as in a gas engine. Sparks can either be made by electric currents, such as welding or spark plugs, or by a flint, same as in a cigarette lighter.

Now that we have learned what can cause an explosion and how it happens, let's examine to see if the same conditions can be found underneath the ground, where earthquakes happen.

Part 3 Underground explosions

Different types of explosions may occur deep inside the ground, but the most obvious one is what needs to be examined.

I would like to mention that since we are not going to study geology or oil drilling, therefore many areas of these fields will not be discussed.

At the old age, when earth was forming, and different layers were rolling over each other, in some parts, earth was made solid. But at some other part, due to many reasons, hollow spaces were also formed among the rolling layers.

This hollow space, can really be hollow, and have just some air, but some others may contain objects, as jungles, alive organisms, buried in them, with plenty of water.

Now after millions of years have gone by, the hydrocarbon materials have turned into fossil fuel, such as oil or methane gas.

This concept has been proven, since when drillings are made in oil fields for this purpose, it reaches fossil fuel.

The size of these hollow spaces can be up to a volume which can hold billions of barrels of crude oil, or cubic meters of natural gas.

I have seen an oil well, which was producing oil under high pressures, for fifty years through a one foot diameter pipe.

As it is indicated so far, a good demolition material such as methane gas, has existed in the ground, and its high pressure, indicates, that the fuel gas is in a very tight and closed container, that can be a good demolition pit with a high coefficient of explosion.

But for an explosion to occur some kind of a ignition is needed.

As described, there are two methods to ignite fuel gas, one like the diesel engine, where the gas is compressed, causing it's temperature to rise and consequently to ignite.

For this method of ignition to take place underneath the ground, it can be assumed that the degeneration of organisms and or hydrocarbon objects produce more and more gas, and since this happens in tightly closed spaces, the pressure causes the temperature to rise to a point where, like the diesel engine, it will ignite the gas.

The second method of ignition is by making some sparks. Let's see how some sparks can occur.

Please refer to figure 3 on page 24. It may look like a pirates treasure map, but let us assume that it is an underground cave, filled with fossil fuel such as

Methane, under high pressure, same as all oil mines are, and a right amount of air for explosions.

What would it take to start a spark in such conditions? It is quit conceivable to have two loose rocks collide together in such a space.

As it is shown in fig. 3, a rock such as RH may fall on another rock such as SH located on the cave's floor, causing sparks and heat.

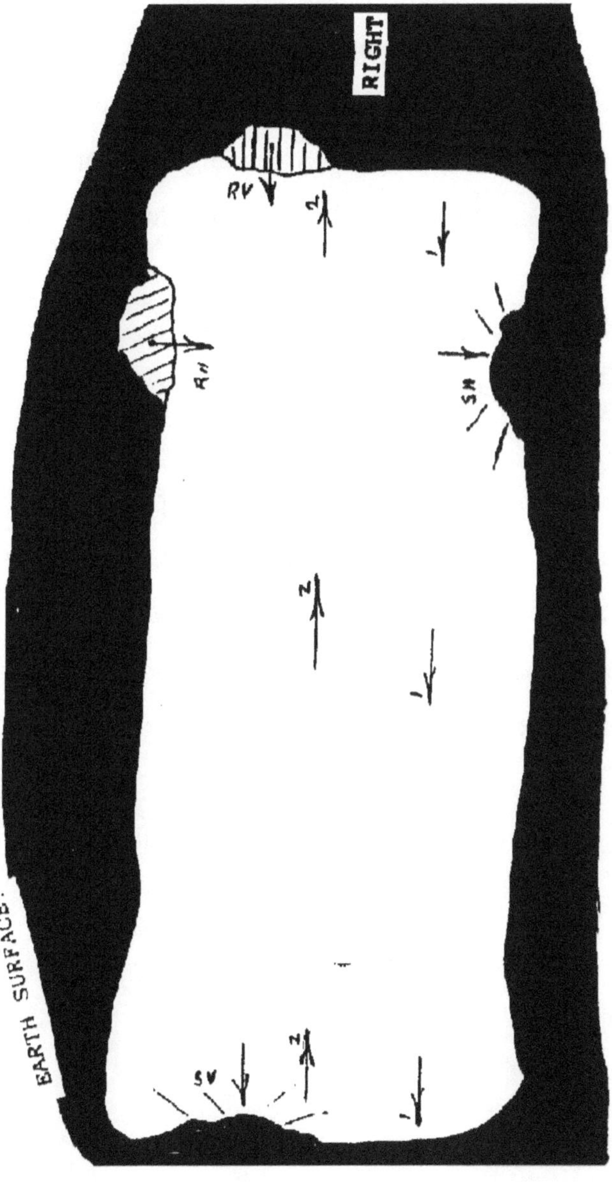

EARTH SURFACE.

RIGHT

FIG. 3 AN UNDERGROUND CAVE.

25

Since all the necessary ingredients to have an explosion are present, Material, tight chambers and sparks, an explosion can happen similar to the one in a gas engine, or a closed room filled with gas to which a spark is introduced.

Fig. 3 is illustrated in such a simple way, so that it can facilitate the understanding of the concept of a closed cave.

In reality there may be many caves that open into each other, never the less, if a rock falls in any one of these chambers and causes a spark, the resulting explosion will have a different configuration from what is portrayed in fig. 3. However the explosion will shake the ground.

Now let us look at what happens when gas explodes.

Part 4 Analyzing The Explosion

Explosion waves have Omni directions, so their movement
are in all directions. Soon after the explosion waves reaches
the right end of the cave, as shown in fig. 3, they will stop.
But since the direction toward the left side of the cave is
free, they will continue moving toward the left.

This movement creates inertia (which is explained later) for
the gas, and also a vacuum at the rear of the movement,
which is the right end of the cave.

When the waves reaches the left side of the cave, they will
stop and start to condense, and build up pressure. At this
point, the right end will have a lower pressure, so the
explosion wave will bounce back towards the right end, and
so on. This "bouncing" will shake the ground, along with
the explosion waves moving back and forth. This is what
we call an EARTHQUAKE.

If the cave is parallel to the earth's surface, there will be a
horizontal movement of the ground, even though the cave

may be at an angle, compared to the surface.

There are cases where the earthquake has a vertical movement, which can be called Vertical Earthquake. This happened in a city near Los Angeles, and as it was told, the ground was moving up about 2 feet.

To see how this is possible turn the fig. 3, about 90 degrees to the left., so that the cave's right end now is the top and the left end now is the floor. What is mentioned above about such an earthquake is true in this case, with the understanding that the waves caused by the explosion will move up and down. So with respect to the ground's surface, the shaking will be vertical. These types of earthquakes can be very destructive.

• A Shake causing an Earthquake.

As there can be hollow space under the ground, of 2 billion barrel capacity, meaning at some parts they can be hundreds of foot high, and when a laundry machine size rock, weighing 5 tons, then assuming 300 tons of rock drop through a 250 feet height, is not an exaggeration.

When this rock crashes down to the cave floor, about 6½ billion foot/pound of kinetic energy will be released. Such an energy can cause an earthquake.

Once I witnessed a refrigerator, weighing about 100 pounds, falling down from the fourth floor of a building, while taking all necessary precautions. All the tenants ran out asking if there had been an earthquake. Judging by this, the assumption that the release of billions of foot/pound of energy can cause an earthquake is correct.

• How powerful can the explosion be?

When gas, in a closed room of about 700 cubic feet, is exploded, some parts of the building will be destroyed. Or a few cubic feet of gas can run a truck for miles at a good speed. Therefore the explosion of millions of cubic feet of gas, in a tight space is imaginable.

• How long can the "Bouncing" last.

The duration of an earthquake is the same as the length of time the explosion waves are bouncing.

1) It could be a moment, an explosion happens, makes a shock, then it fades away.

2) In cases where the fuel gas does not have a high pressure, the explosion wave will bounce for a while and then reach equilibrium, so the cave will be filled with hot pressurized gas, a good source of energy if we can get to it.

3) In cases, where earthquakes last for a minute or more, it provides good evidence that the earthquake is not caused by the released friction between rocks, but that the fuel storage or cave is very tight, and impermeable' due to the surrounding limestone. Therefore the explosion waves will keep on bouncing and the gas molecules will collide with each other creating more heat and pressure. In other words, the explosion waves will be amplified until they can break through a weak point, which would most likely be the top part of the cave, by creating cracks in the walls, then the pressure will be discharged, and the earthquake shakes will

end.

- The release of energy.

When energy is released, it can be converted to three forms of work as heat, sound and movement.

A good example is a gas engine. The combustion of gas in the cylinder generates heat, that is why a cooling system is required and that the exhaust pipe is hot.

It makes a sound, some motorcycle exhaust can show how loud the sound can be.

A moving truck indicates what pressure forces the piston down to rotate the crank shaft and power system.

This is what happens in case of an earthquake.

It shakes a vast area of rugged land, the sound can be heard, miles away from the epicenter. The sound wave pattern can indicate useful information about the explosion.

And there must be some radiated heat.

Therefore when research is done on earthquakes, a heat seeking device should be used to detect any increase in temperature around the quake's focus area, although there can be many obstacles for this kind of a measurement.

CHAPTER

4

VOLCANO

Part 1 Volcano and its Concept

According to the present concept for earthquake and volcano, these two are entirely different phenomenon, one the earthquake is caused by the fault which is a crack on the ground, and the other one, volcano, is caused by MAGMA, a semi melted material, located at a depth of about 150 miles underneath the ground when it moves up, reaches the ground surface and blows out.

So it seems that earthquake and volcano each has different source for it to happen.

Now let us see what is a volcano and what is the concept about it.

A volcano consist of two components, LAVA, which can be a flood of melted rocks and mineral materials, pouring out with a very low viscosity due to its high temperature and flowing down ward on the ground, burning, and destroying everything in its way.

The other component, is a huge amount of gases such as carbon dioxide, sulphur gas, and some ashes, which blow out from a volcano, spread over an area, and may cause some casualties.

The present theory for volcano states that the MAGMA which are at the asthenosphere layer of earth at a depth of about 150 miles, and has a temperature close to melting point,(1200 degrees) when moves up toward the ground surface through the cracks and reaches the ground causes a volcano.

Comparing the volcano with its theory of magma ascending through cracks of more than a hundred miles high rocks it seems that can not be right, and there will be many unknowns as:

- How long will it take for the MAGMA to reach the ground

- What would be the source of energy making MAGMA to have a low viscosity and high temperature as LAVA.

- Moving through the cracks for more than hundred miles can never reach to a point at ground, it will mould some where along such a long distance.

Now we know, what is a volcano, what is its present theory, and some questions about it, so let us make some studies about these cases.

Part 2 What Can Cause Volcano

A volcano happens after an earthquake and loud sound due to under ground explosion. So it seems that both earthquake and volcano may have same source.

Now let us review the new theory about earthquake. With refer to fig. 3 at page 24 when an under ground explosion happens in a cave containing under pressure fuel gas, the explosion wave will bounce in the cave for a while, and finally break through a weak point. This is an earthquake.

It is mentioned that when some potential energy is released, it converts into three forms of kinetic energy as movement, sound and heat.

Movement (which is same as pressure) and sound are explained, now let us discuss the generated heat in the cave due to explosion.

As most of the released energy will be converted into heat, so as the cave is well insulated that there is no heat lost then the cave will be as a furnace with such a temperature that melts every thing in it.

When this explosion happens at a depth under the ground, heat will melt all the surroundings, while the explosion wave are bouncing, and so a weak point will appear and the pressure will be released through that weak point.

Now if this explosion is close to the ground surface, naturally the weakest point will be toward the ground and, so that is where the pressure will push through and blow out over the ground surface and all the melted material will flow on the ground as lava, and the gas will spread into air.

A volcano gas can be consist of different type of gas, carbon dioxide, sulphur gas, and some time ash and others, which is due to impurities in the crude oil as sulfate, salt and others.

The effect of the high pressure and temperature on the impurities will not be discussed, although the results has some information for our subject.

The possibility of gas and melted material to blow out into the earth surface had a similar case as I witnessed.

About 40 years ago at a rugged area not too far from a city, suddenly a column of crude oil, more then a hundred meter burst into the air.

Soon many small lakes of crude oil formed, and as it was summer, oil fume spread far for many kilometers. As it was possible that any spark can make a ball of fire of kilometers diameter, some army units were sent to the area, to move out the people, and no one could even walk in the area.

Fortunately after about a month, the oil column fades down, it was assumed that a rock had jammed the exit vent.

So if the pressure of an oil well can break through and blow out on the ground surface, then the explosion of a fuel storage under the ground will do that as well.

Now, how and when a volcano will be ended is an easy question, but when will it be active again. is hard to answer.

As it depends to the amount of left over pressure and heat at the cave and how the exit channel is jammed, and some other condition.

There are many things that look different, a pot of melted metal and a piece of dry ice, but both are effected by the same subject, heat.

There are some hundreds volcanoes now on the earth, but no new one has happened, one reason, can be as all the fuel caves close to the ground surface have been used.

Earthquake are happening less and less every year. Hopefully in a near geological future, about a million years, there would not be any volcano's or earthquakes on our earth.

CHAPTER

5

EXPLAINING
UNCLEAR CASES

There are some cases in earthquake, which are not explained by the existing theory of earthquakes, but by referring to this new theory, all these questions are cleared and explained.

By referring to this new theory it will be cleared why an earthquake will last for some time and why it shakes the ground. At this chapter more uncleared cases will be explained.

Part 1 What is an After Shock

The release of pressure between rocks which have stopped moving, at a rate of an inch a year, can not make cause for minor earthquakes, which may happen for some months after an earthquake.   About 9,000 minor earthquakes followed the Northridge, California earthquake that happened in January 1994 during ten months.

According to this new theory, as described in Fig. 3, the bouncing of explosion waves will cause the area around it to shake, and may cause some rocks in the underground hollow caves become loose or cracked and then on due to many reason such as gravity, water flow or shocks, they may fall to the bottom of the cave and cause an earthquake or may not be even sensible?

<u>Part 2 Why earthquake is unpredictable</u>

The present theory of earthquakes believes that when the plate movement stops, due to friction between rocks, a pressure will build up to a level, where beyond that, the rocks crush over each other, and the pressure will be released.

This released pressure will be the earthquake.

As the plate movement is traced, so it can be noticed when ever the plate movement is stopped, then after a few decade an earthquake, and the line, where it may happen some where along it can be anticipated. None of the earthquakes that happened were predicted, and none of the predicted earthquakes happened. But using this new theory, in order to predict if an earthquake will happen, the following information is needed.

- To locate all underground hollow spaces or caves

- To find out if there is inflammable and under

pressure gas in the caves.

- To detect I the amount of oxygen needed or ignition is mixed with the gas

- Can a fallen rock make sparks or will it fall in mud and water

So with the present knowledge about earthquakes and the available technical devices it seems that We are not able to predict any underground explosions which can cause an earthquake.

Some time ago it was announced that in China a predicted earthquake occurred. But the procedure is unknown. This is the only case that an earthquake was predicted.

In Japan, by measuring a change in the ground cured surface, an earthquake was predicted. But again out of many earthquakes that happened in Japan, this was the only one which was predicted.

Now let me make a prediction about an earthquake. In a battle field₁ when you are under enemy artillery fire, the safest place is where an artillery round has been impacted a few minutes ago, since it is impossible for two rounds to hit the same spot upon firing.

An earthquake is the same, it is not known if two strong earthquakes ever happened in the same area even after many years.

This is due to the fact that the source of energy required to make an earthquake had been expended. So the people who live in an area where an earthquake has happened once, can be sure that it will never happen again. Just some minor earthquakes or after shocks as described.

<u>Part 3 What is a Fault</u>

A few inches of wide gap or cracks made in the ground is called a FAULT. It usually appears after an earthquake, and is comparable to a crack on a building when it's weight and balance are disturbed.

Which can be due to an earthquake or if the building foundation is broken and collapsed.

The ground and what is on it can be compared to a building, and the underground layers, to it's foundation. So a crack or a fault in the ground can indicate a disturbance in the weight and balance in the area.

Among the reasons for this unbalance, are earthquakes or if the layers of earth under the ground are broken due to flowing waters, etc.

I am not a writer, and I'm not going to write a book but as will be noticed, trying to ease the panic of earthquake among some people, which mostly are elder-lies, and to save those people who are announced missing, might be alive but injured among the rubles demolished by the earthquake, before they are cleaned and dropped in a remote valley.

Also more research needs to be made about earthquakes through this new concept, which is very logical.

Hopefully one day earthquakes can be predicted, as are tornado's and storms.

Until that day, whatever the cause for an earthquake may be, let us see how and why an earthquake demolishes our constructions.

Part 4 Earthquake Insurance Rates

The reason some insurance companies, with all the fraud and high cost never show losses, instead have trillions in assets, is the fact that their figures and policies are based on extreme safe sides of statistics and probabilities.

It was very interesting to me seeing an insurance agent talking about earthquake insurance rates, on T.V. I thought that the insurance industry with their capabilities must have some findings for their rates on earthquake insurance. But I soon realized that I was wrong. As he pointed to a building on a street, he said that the rate of insurance for this particular building is higher than the other building which was two blocks away, since this one is built on a fault.

The fault which he was referring, was an imaginary line running through, not a crack or a gap in the ground. And there were many buildings and streets around this imaginary fault.

We all know that when an earthquake happens, it will shake

an area, not a line and may demolish buildings for many miles from it's epicenter.

Since there is no evidence that a fault can cause a quake, So a different amount for insurance rate for buildings on a fault then the buildings a few blocks away does not make sense.

The different rates should be based on other aspects, not faults.

CHAPTER

6

HOW IT
DESTROYS

<u>Part 1 Inertia</u>

So far, it has been explained what may cause an earthquake and why it shakes the ground. Now let us find out, how an earthquake may demolish a construction.

To find out why it demolish, first we should know a subject called INERTIA.

In nature objects have a tendency to continue their condition with respect to movement. Whether it is Resting or Moving.

To change the status of an object, to stop when it is in moving mode or to move when it is at rest, the object will resist to begin with and need a force to do this job. The amount of force is related to the object's weight and speed.

A good example would be if you tried to move a car when it is parked, with engine and hand brake off meaning to change it's status from rest to move, it takes time and energy, and will not move right away.

Now if the car is moving, and you want to stop it, which means changing it's status from move to rest, it will still take time and energy.

In industry, inertia has many uses and is very effective, whether it relates to large cases such as fly wheels, weighing tons, or fuel gas flowing from carburetor to the cylinder, which weighs a small fraction of an ounce.

Therefore in order for any object to start moving or to stop from moving, it needs time and energy either with high figures as pounds and minutes or low figures as ounce and millie seconds.

There are evidence that shows, we are not happy when we

are born, and no one wants to die either, who knows, may be suffering from birth and death is also due to inertia.

Part 2 Earthquake Shakes

Shakes or vibrations, as they are commonly known, a quick back and forth movement.

This is what an earthquake does, it moves the ground and the objects on it back and forth, which means changing their status from a rest to a moving state, and from move to rest for as long as the quake lasts. Fig. 4, on page 59 shows simply a column or support of a building, which is sitting on top of it's foundation in the ground, and a load such as, a ceiling or a freeway, on its top.

When a horizontal earthquake occurs, the base of the column, as is on the foundation, will move along with the earthquake's shakes. The top of the column is suppose to move with the base, and exactly follow its movement, but due to inertia, this does not happen. The top does not start to move or stop at the same moment with base, but it takes more time to come to rest or to move, so it will have a longer length of displacement and time.

The difference in time and displacement between the top and base of the column, depends on the weight of the load set on the top.

The difference in movement is quite obvious, in a high rise buildings when earthquake happens, the top floors will move much more from side to side, than the basement.

It looks that the ancient architects of many thousand years ago had solved this problem, as will be described later.

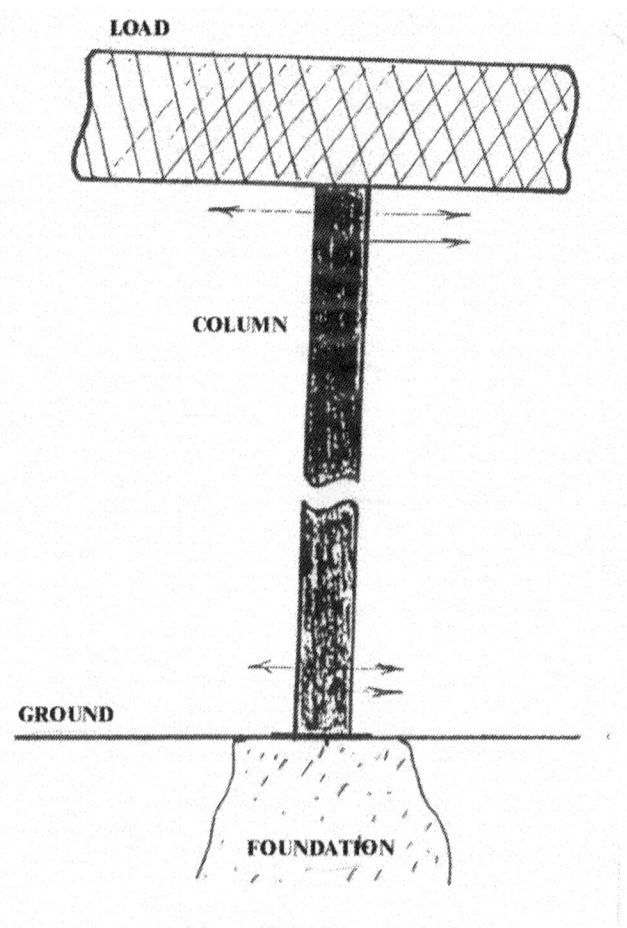

FIG. 4

In a building there are many columns which all are joined together, but this can not cancel the effect of inertia, as it is a natural phenomenon such as gravity. In case of a vertical earthquake, the effects of the shock will be the same, but in a different direction, therefore this should be figured out a different view.

FIG. 5   TWO MODLE OF MUD MADE BUILDINGS.

CHAPTER

7

BUILDINGS
AND EARTHQUAKES

Part 1 Mud made house

As we are not going to study the civil engineering, so many subjects about building constructions will not be discussed here.

What will be studied is the effect of earthquake on buildings with respect to the construction materials and some physics and mechanic principals. The reports that are given following an earthquake vary quite a bit. One might hear that a 5 degree earthquake destroyed 8,000 homes, left 4,000 people dead and 500 missing, but in a different area, a 6.5 degree earthquake left some damages, and a few casualties.

Both news are sad, but the reason of such a difference in the number of casualties compared to the even a lower degree of the quakes is due to the type of buildings in the areas of the earthquake.

I believe that a poor is unlucky too, but still let us have a look at a typical building which is most commonly damaged during an earthquake.

Fig. 5, on page 60 shows a house which it is just a room, and can not be wider than 15 feet, this house made from a mixture of mud and straws.

In one of the models, the ceiling is made of tree trunks, and some branchlets covered by mud and the other one has an arched ceiling, and both   will collapse easily when is shaken.

Once I asked a villager why they build such type ceilings; it takes more material and time to built. He answered that we have a lot of time and mud, why should we buy a tree !

Mud has an elasticity of almost zero, and doesn't hold the building firmly together. During the first few shocks of an earthquake, the building turns into a pile of dirt. If they are far from the epicenter, the building will crack, and will need to be repaired, or rebuilt, which is the same thing except there aren't any casualties.

These type of buildings are no more built at cities, but can be found at some remote village of some middle eastern countries.

Part 2   Cement made buildings

Buildings which are made of cement and bricks, are held firmly together and the weight of the material makes the walls more stable.

As is shown in fig. 6B, in this type building, the walls are thick, up to a few feet, which is a good technique for resisting against earthquakes.

However, since the material that is used are heavy, so the accumulated force due to shakes, which is related to the weight, will be high and may cause damage.

The direction of the earthquake shakes is very effective. As shown by flash in fig. 6, if the shake is parallel with the walls, it will be pushed to it's end which will hold it from falling apart, and may only crack. But if the movement is perpendicular to the wall, as it is pushed towards the open area, the walls may collapse down.

In some old castles, the thickness of the basement walls are more than a yard. It is possible that the architects, in those days new a relation between the thickness and the height of a wall.

Who knows, may be one of the reasons that the great wall of China is so thick, could be so that it would resist against earthquakes too.

Part 3 Steel Made Structures

At the present time, buildings have a steel structure. The supporting iron beams all are joined together, so the entire building can be considered as a big iron solid box. That is why earthquake has not damage a high rise building.

Earthquake shakes, can not open a nut and bolt or sheer off a steel plate, which is 1/2 an inch thick.

The iron beams are selected of a size, that will be strong enough for the earthquake shakes.

At new buildings, light construction material, as dry wall, lumber, etc. are used so the accumulated force due to earthquake shakes are not strong to break a lumber or cut a 1/4 of an inch iron nail.

About the ceiling, in case of a horizontal earthquake, as the shakes are parallel to the ceiling, it would not be pushed to collapse or crack, but a vertical earthquake, as the ceiling is pushed up and down, it may cause to be damaged, how ever it is made by light material too.

Obviously there are many types of constructions, but in all cases, the effects of inertia and earthquake can be applied to them as well.

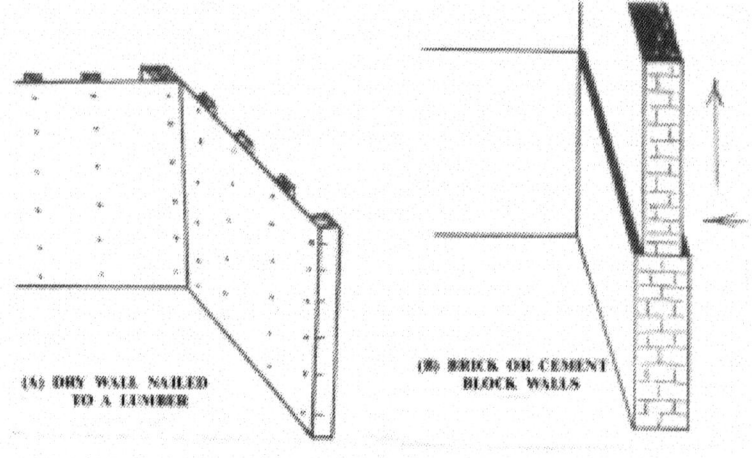

(A) DRY WALL NAILED
TO A LUMBER

(B) BRICK OR CEMENT
BLOCK WALLS

CHAPTER

8

ANCIENT BUILDINGS

Part 1 The Pyramid

At the previous chapters, we noticed how buildings are effected by earthquake, but those are the buildings, that now are used to, live, and as mentioned, there can be other types of buildings too.

Now let us have a look at a building, which was made thousands of years ago, and find out how an earthquake can effect it.

This is a stone made building, and has a pyramid shape. The most famous ones are the three pyramids in Egypt. But in other parts of the world, like Mexico, the Mayan temples are built on a topless pyramid, or in Iran, the tomb of an ancient king, who is also believed to be prophet of Israel, by the name of CYRUS, is also built on a high, topless pyramid.

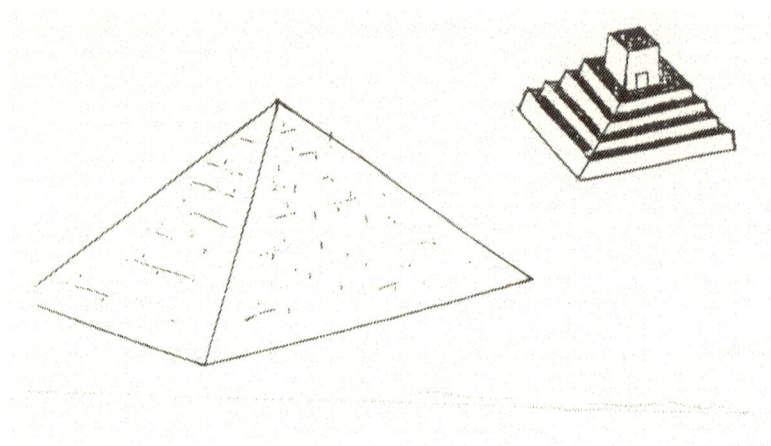

FIG. 7 PYRAMID AND KING CYRUS TOM.

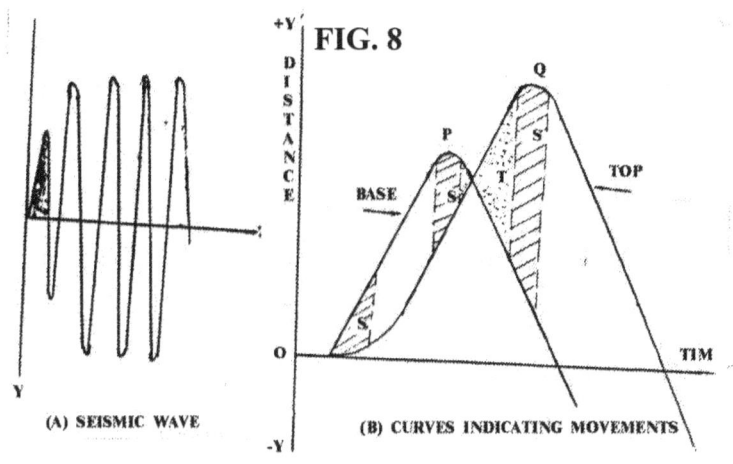

FIG. 8

(A) SEISMIC WAVE   (B) CURVES INDICATING MOVEMENTS

There must have been good scientists in Egypt, in those days. As an example a man by the name of Hero established a formula to calculate the surface of a triangle by having the length of each sides of it. This formula is still used, but is proven by using a different method and is called Hero formula.

Fig. 7, on page 72 shows a pyramid, and the tomb of the ancient king of Iran CYRUS which is very similar to the Mayan Temple in Mexico.

It is not known why Egyptian Pyramids, which are Pharaoh's tombs are made in this shape, or why Mayans in

69

Mexico had built their temples, on topless pyramid. But the reason CYRUS' Tomb was made in this shape is known, his wish was to "Bury me on the ceiling".

The Mayan Temples in Mexico and CYRUS' Tomb in Persepolis, Iran and the pyramids in Egypt all are located in earthquakes zones. Therefor I believe, that they must had some knowledge of physics subjects as inertia, that these buildings were made in the shape of pyramids so that they would resist earthquakes.

Now if this idea is correct then the ancient architects were right to make a building in pyramid shape. As will be discussed in later chapters.

About the Egypt pyramids, who knows, may be the Egyptian Pharaoh, some day told his architects, "Build me a tomb, that nothing may disturb peace on my body after my death."

So pyramid is what the Egyptian architects figured out on their search.

Pyramid, this ceilingless building, contains many rooms at different levels, some rooms are so big that pharaoh's ship is placed in it.

The building makers must had a good knowledge of weight and balance that could build, so many rooms in such an odd looking construction.

This nice symmetrical looking building, indicates that the makers, had good technical knowledge, and were using some scales for measurements that were able to make it.

To find out how a pyramid will be effected by of an earthquake, we should see the details of an earthquake shake effecting on a building support and then apply it to a pyramid.

Fig. 4, on page 59, shows a column supporting a load on its top and sat on a foundation in the ground. When earthquake happens the column's base moves from its center to left and right but the top part, which is carrying a heavy load, due to inertia, will not move at the same moment. As the base moves, and also when the column's base stops to reverse its direction the top continue its move and stops later to reverse its direction.

This means that top part of the column moves through a longer distance and time then the base.

In order to demonstrate this difference in movement, which can be on a solid iron beam let us draw a graph to illustrate time verses the displacement.

Fig. 8 on page 37 shows two graphs, A and B on both graphs line OX indicates time, and line $0 \pm Y$ indicates the distance where the column moves left and right from its center.

Graph A is the earthquake shakes registered by a seismograph, a half a cycle shaded curve, is what will be used.

Graph B has two curves, one is <u>base</u> which indicates the column's base movement, and it moves with the same pattern as the earthquake shakes. The other curve top indicates the column's top movement.

Both curves are enlarged, so that the details can be better noticed.

When the base starts to move from point B through a distance to point P and stops to reverses direction to return to its' original position, and continue on with the earthquake's shakes. The top curve will not move at the same moment, from point B, due to inertia, but will move later. Also, at this time on point P, when the <u>base</u> has stopped to reverse its' direction, top curve which indicate the column's top movement will continue moving on to point Q, where it will stop to reverse its direction, in order to follow the bases movement, since base is it's primary move.

By studying the difference in time and displacement of these two curves, which actually are the movement of the column's top and base the followings are found:

- Although the curve top is originated to move by curve base but it moves through a longer time and distance, which means the curve top which is actually the earthquake wave is amplified.

- The shaded sections S of the graph indicate that as one end of the column is moving, but the other end its stopped. For that period of time the column is under stress.

- The dotted sections (T) of the graph indicate that two forces, called (TORQE) in different directions are applied to the top and base of the column, causing, it to rotate as can be noticed from the graph.

The mentioned condition is what is applied to the column for every half a cycle so twice for every shake, and will continue for the entire duration of the earthquake.

The graph descriptions about different points on the curves, are theory, as time can be at range of milliseconds. But when an earthquake happens, and a building shakes, the difference in movement at base and the top floors are noticeable.

Based on this the above theory is true and applicable.

Part 2 Pyramids and Earthquakes

In the previous chapter, the theory and practical effects of an earthquake's shakes on buildings was studied, and find out that what makes an earthquake destructive is the weight of the top part of the building. Which acts as a load and amplifies the earthquake shakes, and makes the destruction.

So if some how we can eliminate this load, earthquake will not be destructive.

Now a days, buildings have steel structures, that's why the shaking due to earthquake are not a problem. But in the old days where the best building materials were stones and pine trees earthquake's were a disaster.

I believe that pyramid shape for high rise buildings, at many thousands years ago, can be an evidence that the architect realized the problem of the weight of the upper part of a building during an earthquake, and how to omit it was a puzzle, which by making a building as pyramid shape, it was solved by the best possible way.

A pyramid shape, not only, the upper load of a building, is omitted, but also the building center of gravity is lowered down close to the ground.

Any object has a center of gravity that in any movement, such as earthquake shakes, effects the weight and balance of that object, that is why the race cars are almost flat on the ground, to not turn over easily.

I believe that King CYRUS' palace in Persepolis was made by Greek architects, who weren't slaves and been paid.

The pyramids in Egypt have a flat surface, sides and are complete. But CYRUS' Tomb or the Mayan Temples are a top less pyramid and goes up step by step each level is being smaller than the previous one.

This may indicate that the Greeks and the Mayans architects had a different concept about their pyramids, then Egyptians, and may had some idea about integral in mathematics as each level of a pyramid is less in size and weight then its lower level, consequently it does not produce a force or move more then its lower level. So no destruction happens.

Many stone castles made in ancient days, now nothing is left except some remaining, but the pyramids are still standing.

CHAPTER

9

MEASURING
EARTHQUAKES

Once an earthquake has occurred, it takes a while to find out it's magnitude. This is obtained from a table, established by a geologist name Richter, and that is why the units of the table are called Richter. These units have logarithmic values. The log base is not said but it is known that the figures have a difference of thirty units, meaning an earthquake of 3 magnitude is thirty units stronger than a 2 and sixty units stronger than a 1.

These are just numbers and does not specify anything.

In order to measure something, all of its specifications should be found out, to known what to be measured.

"An earthquake is a natural phenomenon, when it happens, it shakes the ground for a while and it ends". Let us analyze this sentence. "Shakes for a while". There we can notice three components in this statement.

- While: Meaning time; how long an earthquake lasts.

- Shakes: A back and forth movement is what we call shake.

- Move: Which can have a unite length indicating the displacement.

Therefore an earthquake has three components, time shaking, and length of each shake which is displacement.

As we notice each of these components, has it's own effect and characteristics and unit of measurement. There can not be any mathematical configuration which can indicate these three components by one figure.

As mentioned each of these component has its own effect. So if any one of them be high even if the other two are low, still can make some damage. Mathematically, the length of a shake and the number of shakes per second can be changed into speed as in some calculations, the speed is needed and is figured out by its square value. But speed won't tell us what we need.

If 2 feet move in one tenth of a second, it will have a speed of 20 feet per second. The same result will be obtained if we have 4 feet moving in one filth of a second. But in case of an earthquake these two speeds have a different effect.

Speed doesn't show the length of displacement or the number of "Shakes" per second.
What Is necessary to be known is to measure for how long, how far and how many times per second a building was moving.

The intensity of this size explosion which causes an earthquake can be measured, by comparing it to explosion of equivalent of tons of demolitions as T.N.T. or dynamite which can be up to many 100 million tons.

In case of earthquake, in addition to this method, as the created sound power Bell (B) is directly proportional to the explosion intensity, so by some research, the magnitude of sound wave can be a means to figure the magnitude of an earthquake.

CHAPTER

10

THE PANIC
OF EARTHQUAKE

Part 1 Cause of Panic

An earthquake happens for a short period of time, it may make some damage and casualties or not, and some time later, it is forgotten, but some of our loved ones, who are elderlies, are in panic of earthquake which will effect the family life.

I have met such people, and it was my motivation to write these notes.

What increases the panic, and makes an earthquake so fearful is the exaggeration about it.

And since we are not familiar with earthquakes, so it is known as a terrible.

A T.V. program about earthquake, was full of fearful sights, walls were falling all around the room. Flood from a broken dam was washing away people, a broken high voltage line was dropped down on the ground, making sparks. In fact none of these events can happen due to earthquake.

A dam is far from a city, and is located at a mountainous area, it is built so that it will not be demolished by an earthquake, It may crack but not totally destroy and create floods.

If a high voltage over head line, due to any reason, is dropped down, the safety device will cut off the power In a few seconds.

It is described how earthquake effects a building, shakes would not pull out a nail from lumber, or break a dry wall, and if it happens, I don't know how, only one side of the room may fall down, not all the sides. Now let us see besides rumors what else makes earthquake's seem so terrifying.

Cases that make damage and casualties are frightening, specially when they are not under control, and the sources are not known.

It would be very disturbing to hear loud noises mixed with screams at night, but when the source of these noises is identified as being a teenager's party, then all fear, will subside.

Let me give you another example, it was about 60 years ago that for the first time I saw a car. About 30 people had gathered around it. Suddenly a loud noise rose from the front and the back of the car, the driver was shouting at the same time. Every one, started running away from the car, and I was one of the runners. When I got to a distance, I turned around and started observing the car again.

The reason we were all so scared is the fact that the car was unfamiliar to us. But today no one is afraid of cars even though they make much more casualties than earthquakes.

The first time when tanks appeared at a war front, was during the First World War, by the allied countries against Germany.

As tanks started moving toward the German units, they were disturbed and withdrew for kilometers. But now we have procedures for individual to fight against tanks.

Another reason of fear or panic from an hazardous event, is if the event can not be controlled, such as storm, neither it's intensity, nor it's direction can be controlled.
There are other causes of panic which would be out of context at this point.

<u>Part 2 Vain Panic</u>

Based on the facts given, we know now that there is really no reason to panic from an earthquake. Now we know how it effects a construction, so with the present materials and steel structures used for a building, it will not be demolished by an earthquake, only some minor superficial cracks may occurs.

That is why none of the high rise buildings been damaged by an earthquake. A strong blast of two hundred pounds of explosives, made by a terrorist group in a high rise building in New York, could demolish a limited area of that huge building. So we can be sure that our architects are familiar with earthquakes, and it is under control.

The 700 casualties in the 1906 San Francisco earthquake will not happen again.
It is true that we do not know when and of what magnitude an earthquake will happen, but this should not make us worry or be in panic at all times.

An earthquake, by itself is not harmful. It you're sitting in a car, or park and an earthquake happens, it will not hurt you. Most of damage and casualties are due to the defective buildings, not earthquakes.

I was watching a T.V. program in which someone was saying that the BIG earthquake WILL happen, but when it will happen and where it will happen we do not know.

I have heard this story before, but how come that "where" and "when" is not known, but "will" is known. This incorrect prophecy is what causes panic among people. Even though there is no evidence of a big earthquake to happen. But if it does happen, as long as the building in which we live in are safe, we have nothing to worry about.

When an earthquake happens, and I am separated, feel very light, I can pass through walls and no one can notice me when I talk to them, then it is too late to worry about earthquake.

But if an earthquake happens and ends, and I am alive, why should I have panic, there are many unexpected events threatening my life, should I suffer for probabilities, and ruin sweet life and not enjoy it.

Part 3 Panic Treatment

There are many people in California who are in panic of earthquakes, where earthquakes happen often. I know of a family who moved out of the state to avoid earthquake's. Among these persons the case of a lady I know is noteworthy. As it starts getting dark, this lady would worry about earthquakes, every little sound would make her think.

As she was really suffering, she decided to see a doctor, a neurologist, since she thought it must be a nervous matter. She was given some pills, which she could sleep for a few months, but their effect went away after a while and she had the problem again.

It is questionable why she wasn't directed to a physiologist who would better be able to help ease her panic. California is an earthquake zone, and there are some people who are elderlies, and suffering of such panic, they are not known and they do not know where to go. There is not enough psychologists at medical centers to visit these and other patients. I am hoping that these facts about earthquakes will help to ease the panic of some of them if not all.

CHAPTER

11

MISSING
IN EARTHQUAKE

The earthquake in Mexico City made lots of casualties and many damaged buildings.

As T.V. showed a hotel which was demolished, many had died and more than hundred were missing. The cleaning process began. They started to pick up the collapsed pieces, using heavy machinery, and disposing them to remote valleys.

I was thinking about the missing persons, where else can they be except among these demolished walls, some of them may be dead, but some may still be alive and badly injured, they will die, but some injured persons who have been rescued after twenty days were still alive.

Injured people underneath in the demolished parts is not a matter of may be, especially when there are some missing persons.

A good example, is the case when an earthquake happened in Northridge, California in January of 1994.

The earthquake made much destruction, some people who were searching to find any injured persons under a demolished building, as the next day trucks and loaders, were supposed to come and start cleaning the area.

When they finished their search, and were ready to leave, one of them said I think I heard a noise, so they start looking around again, and found an injured man trapped under the demolished blocks.

This man could be alive after being loaded in a truck, and dropped at a remote valley, finally he would die, but with much more injuries, and what a death.

The T.V. showed when savoir met the saved man in the hospital. They were looking at each other with thoughts and sensations.

In a village which was destroyed by earthquake, a mother was asking the people to help her so that she could get her daughter's body.

She said, I was sleeping between my two daughters when earthquake happened and walls started falling down, I jumped up and grabbed one and pulled her out with me, now how can I leave my daughter's body under the demolished walls.

She hired a few men to clear a ten foot by ten foot area, to find the body. I can't describe the details, when the crashed body of the nine year old girl was found, many people left, most people were crying, and the poor mother blessed the workers, and said now I put her in a place that whenever I want to remember her I know where to go.

To help locate missing people trapped underneath the

collapsed buildings, beside searching, dogs or sound detectors are used, these are the only methods to use, and are helpful, but have weak points. Dogs can not smell people, if the wind is blowing far from the dog, and who knows what goes through a dog's mind if he feels that the person who is located are hard to reach, what would be the dog reaction, ignore or what.

A sound detector, which is actually a sound amplifier, will amplify any sound, whether it's a roach moving or water flowing through pipes which can be stronger than a heartbeat.

So it is better to use other means like an electronic device, as a small beacon, be available to people who live in an area where earthquake can cause damage. So we should be more concerned about how to clear the demolished pieces of construction if there are some missing persons.

CHAPTER

12

PREPARING
FOR AN EARTHQUAKE

We as a human race can be prepare our selves for almost any kind of disaster that threatens us, except one case, which is beyond our discussion.

Fatal cases such as fire, car accident, or even a battle filed, procedures are set to eliminate or minimize the fatality. So an earthquake is no more complicated than these cases.

In case of an earthquake the procedure to reduce or eliminate an earthquake casualties, are fortunately made. Since the casualties have one source, which is construction, so the type of materials and buildings are selected so that earthquake shakes can not damage them.

The following reports were released about some events that

happened after the Northridge earthquake in California.

- A joint in a building structure was supposed to have 4 bolts, but only had 3, one bolt was not used.

- Some parts of a steel plate used in a joint of a building structure was sheered off.

- A defective building was demolished and 16 people died.

So the above cases indicate that the first phase to be prepared for an earthquake before it happens is to have a safe building.

For this purpose, and regarding the mentioned reports above, in an earthquake zone, often building Inspections depending on the material and the date that the building was made is required.

If this procedure was done, at least those 16 people would

be alive now.

Inspections forms can be prepared by the city authorities, but some construction companies should be authorized to do the inspection and the necessary repair, and not to put all the load on the city administration.

Earthquakes happen quickly, therefore most of the advise given is good for when the quake is over.

At the time when the room is shaking and things are falling down, who can remember what to do.

What can be done before an earthquake is to be sure the place where you live is safe and is inspected by the authorities.

- In a large building, remember your location and the shortest way to open air. During those seconds that looks like the whole world is shaking, stand up and be as a leader, and have courage.
- You always make better decisions when you are calm.

END

ABOUT THE AUTHOR

Hired in 1942 by British Petroleum (BP), for 16 years I worked at oil refineries, oil fields, and had a chance to pass a four year course in the company's TECHNICAL INSTITUTE. Then, I joined the army in which most of my service was in the engineering corps. That is why you will see some examples of battlefields in my book. I am a retired colonel after 30 years of service.

-Bahram Hojabri

Questions about this subject?

You can call: (818) 242-9628

Or write: 435 W. Elk Avenue, Glendale, CA   91204